ARTEMIS UNVEILED:

THE MOON ELECTRICAL SHIELD

Uncovering the Secrets of Lunar Space Weather

BERT D. BOUFFARD

GRATITUDE

Dear Reader, I want to give special thanks to you for your efforts and concern towards educating yourself. I want to thank you for choosing my book among other books.

We have put in so much effort so that this book will satisfy your interest in knowing more about "ARTEMIS". Join us today to know everything about ARTEMIS

Copyright © Bert D. Bouffard, 2024

All rights reserved. No part of this publication may be reproduced, distributed, or transmitted in any form or by any means, including photocopying, recording, or other electronic or mechanical methods without the prior written permission of the copyright owner, except in the case of brief quotations embodied in critical reviews and certain other non commercial use permitted by copyright law.

TABLE OF CONTENT

INTRODUCTION TO ARTEMIS AND THEMIS MISSIONS 7
    The Beginning of a Mission 8
    THEMIS: A Mission with Purpose 10
    The Transition to ARTEMIS 12
    A New Era in Lunar Exploration 14

CHAPTER 1 18
    The Genesis of the THEMIS-ARTEMIS Mission 18
    Origins of the THEMIS Mission 18
    Objectives & Goals 21
    The Transition to ARTEMIS 24
    THEMIS-ARTEMIS Mission Overview 27

CHAPTER 2 32
    The THEMIS Satellites 32
    Launch and Deployment 32
    Launch Vehicle and Site 33
    The Function of Swales Aerospace 37
    Collaboration with NASA's Jet Propulsion Laboratory 43

CHAPTER 3 46
The Mission Phases of THEMIS........................... 46
Initial Mission Phase..46
Orbital Adjustments and Objectives........................ 50
Magnetotail Exploration and Data Collection......... 52
Observations on Substorms and Magnetic Reconnection..54

CHAPTER 4 59
Groundbreaking Discoveries and Data....................59
Magnetic Ropes and Solar-Terrestrial Interaction..59
NASA's "30-kilovolt battery" comparison............... 62
The Role of Magnetic Reconnection........................ 65
The Moon's Role in Space Weather........................ 68

CHAPTER 5 71
Transition to ARTEMIS... 71
ARTEMIS Mission Objectives................................. 72
Challenges of Transition..75
ARTEMIS-P1's orbital manoeuvres......................... 79
ARTEMIS-P2's Mission to Lunar Orbit.................. 81

CHAPTER 6 88
The ARTEMIS Lunar Operations............................ 88
ARTEMIS-P1 and ARTEMIS-P2 are in lunar orbit. 89
Understanding Lunar and Solar Interactions......... 93
Exploring the Moon's Exosphere and Magnetotail. 95

CHAPTER 7..**105**
    Collaborative Efforts and International Contributions..105
    Contributions by Canada, Austria, Germany, and France... 106
    The Role of NASA Launch Services Program (LSP).... 113
    Integration and Support from the University of California, Berkeley... 117

CHAPTER 8..**124**
    Future Prospects and Implications........................ 124
    Hidden Dimensions of the Lunar Plasma Environment.. 125
    Lunar Magnetism: A View Into Planetary Evolution.. 128
    The Impact of Lunar Space Weather Studies..........129

APPENDICES..**139**

INTRODUCTION TO ARTEMIS AND THEMIS MISSIONS.

When humans first saw the Moon, it seemed as a distant, quiet sentinel in the night sky—a constant emblem of cosmic beauty and mystery. However, underlying this peaceful exterior lurks a dynamic and chaotic world shaped by cosmic forces. The ARTEMIS and THEMIS missions are daring attempts to lift this cosmic curtain, exposing the Moon's hidden mysteries and interactions with the solar wind and the Earth's magnetic field. In this excursion, we dig into a fascinating area of space science that gets us closer to understanding not just our celestial neighbour, but also the processes that form our whole solar system.

The Beginning of a Mission

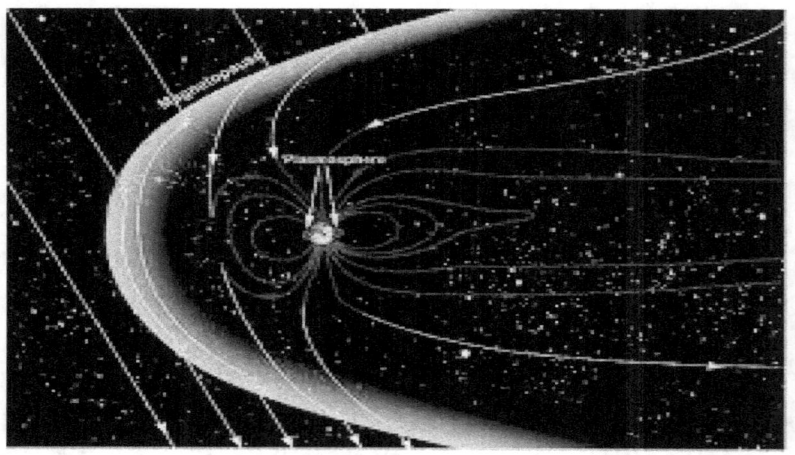

The narrative of ARTEMIS and THEMIS starts with a mission to discover the secrets of Earth's magnetosphere, a large, protective bubble of magnetic fields that shelters our planet from solar radiation. The THEMIS

(Time History of Events and Macroscale Interactions During Substorms) mission, which launched on February 17, 2007, was designed to study the dynamics of the magnetosphere and its complex interactions with solar wind. Five identical satellites, each a wonder of contemporary engineering, set off on a mission to investigate substorms and magnetic reconnection occurrences in the Earth's magnetotail. The mission was intended to track the origins of geomagnetic activity and provide insights into the causes of the stunning auroras that illuminate the Arctic sky.

However, the mission's scope went well beyond the Earth's atmosphere. As the spacecraft passed through the magnetosphere, it became clear that the

Moon—a celestial entity that has long captivated us—was inextricably related to these cosmic occurrences. This conclusion paved the way for an ambitious expansion of the THEMIS mission, resulting in the development of ARTEMIS (Acceleration, Reconnection, Turbulence, and Electrodynamics of the Moon's Interaction with the Sun).

THEMIS: A Mission with Purpose.

Each THEMIS satellite was built with an array of advanced equipment to help researchers understand the complexity of space weather. These equipment, ranging from fluxgate magnetometers that measured magnetic fields to electrostatic analyzers that identified charged particles, presented a detailed image of Earth's surroundings. During the mission's early phase, the satellites operated in a string-of-pearls pattern, providing critical data on geomagnetic substorms and magnetosphere dynamics. This phase was followed by the "Dawn Phase," in which the satellites were moved to investigate the dawn side of the magnetosphere, which is of particular interest because of its unique interactions with solar winds.

One of the mission's major accomplishments was the discovery of magnetic cables linking Earth's upper atmosphere to the Sun. This finding, which confirmed the hypothesis offered by Kristian Birkeland over a century ago, demonstrated the solar-terrestrial electrical interaction in operation. The THEMIS satellites offered a precise glimpse of these interactions, which NASA described as a "30 kilovolt battery in space," with magnetic ropes pouring an incredible 650,000 amps into the Arctic.

The Transition to ARTEMIS

As THEMIS neared the conclusion of its primary mission, the attention moved to the Moon. This transformation resulted in the formation of the ARTEMIS mission, which was entrusted with researching the Moon's unique space environment. In 2010, the spacecraft started a series of lunar flybys and deep space excursions, laying the groundwork for their eventual entry into lunar orbit.

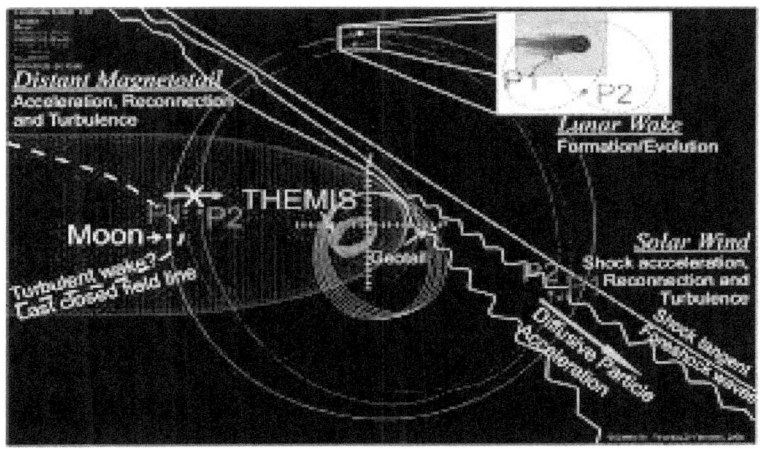

The ARTEMIS satellites, renamed THEMIS-B and THEMIS-C to ARTEMIS-P1 and ARTEMIS-P2, proceeded on a difficult trip to reach lunar orbit. Their arrival was a historic milestone since they were the first spacecraft to circle the Moon's Lagrange points—points of gravitational equilibrium where the forces of Earth and Moon balance, providing a steady vantage point for scientific research.

A New Era in Lunar Exploration

ARTEMIS-P1 reached lunar orbit on July 2, 2011, with ARTEMIS-P2 following on July 17. These spacecraft were not only pioneers in lunar orbit, but also important in understanding the Moon's space weather environment. The ARTEMIS mission was to investigate the Moon's interaction with the solar wind, as well as the effect of the lunar environment on Earth's magnetosphere. This finding was critical for understanding how solar activity impacts the Moon and, in consequence, our Earth.

The ARTEMIS mission was designed to explore phenomena such as the acceleration of charged particles near the Moon and the turbulence created by solar winds. Scientists anticipated that by analyzing these interactions, they may learn more about the Moon's function as a natural laboratory for researching space weather and its impacts on both lunar and terrestrial ecosystems.

Why This Matters

The ARTEMIS and THEMIS missions represent more than simply scientific curiosity; they are critical to our knowledge of space weather and its possible effects on current technologies. Space weather occurrences may interrupt satellite communications, GPS systems, and power grids, hence it is critical to understand the underlying mechanics that cause these events.

By revealing the mysteries of lunar space weather, the ARTEMIS mission not only expands our understanding of the Moon, but also supplies important data that may help protect our electronic infrastructure. The information gathered from these missions help us prepare for space weather events and pave the path for future exploration and discovery.

A journey into the unknown.

As we go through the pages of "ARTEMIS Unveiled: The Moon's Electric Shield," we welcome you to discover the intriguing realm of lunar space weather. From the inner workings of the THEMIS satellites to the startling findings of the ARTEMIS mission, this book provides a complete and compelling narrative of our search to comprehend the forces that shape our cosmic neighborhood.

Discovering the mysteries of the Moon's electric shield not only illuminates the dark regions of space, but also provides a better understanding of the dynamic interaction of cosmic forces that impact our globe. Join us as we explore this astonishing chapter in space exploration and learn about the incredible tales behind the ARTEMIS and THEMIS missions' data and findings.

CHAPTER 1

The Genesis of the THEMIS-ARTEMIS Mission

Origins of the THEMIS Mission

In the wide expanse of space, where cosmic forces orchestrate a symphony of charged particles and magnetic fields, the Earth's magnetosphere acts as a quiet guard. To investigate this cosmic shield, NASA launched the THEMIS mission on February

17, 2007, a ground-breaking effort. But, before we get into the mission details, let us first look at the genesis and grandiose concept of THEMIS.

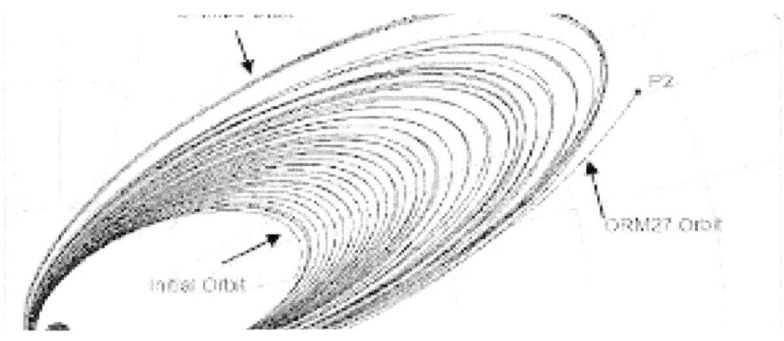

The Time History of Events and Macroscale Interactions During Substorms (THEMIS) mission was designed to address a long-standing space scientific question: what causes geomagnetic substorms and magnetic reconnection. These events, which cause the spectacular auroras seen in northern sky, are inextricably tied to the

Earth's magnetosphere and its interaction with solar winds.

THEMIS originated in the early 2000s, when researchers identified the necessity for a specialized mission to investigate the magnetosphere's dynamic dynamics. They aimed to answer crucial questions: What causes substorms? How does a magnetic reconnection event unfold? How do these processes impact space weather, and by extension, Earth's technological systems?

The THEMIS project was precisely designed to launch a constellation of five satellites, each equipped with cutting-edge instrumentation to offer a complete picture of these occurrences. The objective was to record the complicated dance of particles and fields in the magnetosphere, providing unique insights into the systems that govern space weather.

Objectives & Goals

The THEMIS mission was more than just a scientific curiosity; it was a quest with specific goals aimed at solving the mysteries of Earth's magnetosphere. At its heart, the mission aimed to answer several crucial questions:

1. Understanding Substorms: Substorms are severe disruptions in the Earth's magnetosphere that cause vivid auroras and may disrupt satellite communications and power networks.

THEMIS sought to understand the origins of these substorms by tracking their progression and identifying the primary reasons driving their emergence.

2. Magnetic Reconnection: One of the primary goals of THEMIS was to investigate magnetic reconnection, which occurs when magnetic field lines from the Sun and Earth rejoin, releasing energy that accelerates particles and triggers space weather occurrences. THEMIS's goal in exploring this mechanism was to improve our knowledge of how solar activity affects the Earth's space environment.

3. Spatial and Temporal Dynamics: The mission was intended to record the spatial and temporal fluctuations of magnetic fields and plasma in the magnetosphere. THEMIS may detect these processes from many perspectives, resulting in a three-dimensional picture of magnetospheric activity.

4. Technological Advancements: In addition to its scientific objectives, THEMIS sought to improve space-based equipment and observational methodologies. The mission's success would pave the path for future missions while also improving our capacity to monitor and anticipate space weather phenomena.

The THEMIS mission's methodology required a complex array of equipment on each satellite. Fluxgate magnetometers measured magnetic fields, electrostatic analyzers detected charged particles, and solid-state telescopes saw high-energy particles. The combined data from these devices was supposed to provide a thorough and nuanced knowledge of the magnetosphere's dynamics.

The Transition to ARTEMIS

As the THEMIS mission advanced, a new frontier emerged—one that reached beyond Earth and into the domain of lunar research. This change signaled the start of the ARTEMIS mission, a continuation of the THEMIS goals with a new emphasis on the Moon.

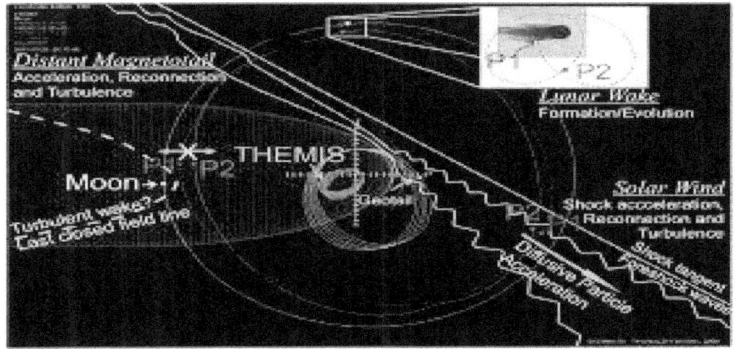

Renaming and Reconfiguration

The switch from THEMIS to ARTEMIS was more than simply a name change; it marked a rethinking of the mission's objectives and a shift in emphasis. While THEMIS focused on the Earth's magnetosphere, ARTEMIS (Acceleration, Reconnection, Turbulence, and Electrodynamics of the Moon's Interaction with the Sun) was created to study the Moon's interaction with the solar wind and how it affects the Earth-Moon system.

A variety of reasons influenced the decision to rename the mission and modify its goals. As the THEMIS satellites collected data, scientists discovered that the Moon's unique location in orbit provided an excellent vantage point for investigating space weather events. The Moon's lack of an atmosphere and magnetic field makes it an excellent place for studying the interactions between solar wind and the Earth's magnetosphere.

The reconfiguration included changing the mission's equipment and spacecraft operations to concentrate on lunar research. ARTEMIS was designed to do lunar flybys and enter orbit around the Moon, opening up new chances to research how solar wind influences the Moon and how the Moon's space environment affects the Earth's magnetosphere.

THEMIS-ARTEMIS Mission Overview

The THEMIS-ARTEMIS project exemplifies how scientific discovery grows and responds to new findings. The inaugural THEMIS mission was successful in its purpose of investigating geomagnetic substorms and magnetic reconnection. However, the potential to study the Moon opened up an exciting new chapter in space research.

ARTEMIS, together with its two satellites (previously THEMIS-B and THEMIS-C), launched a new mission to investigate the Moon's space environment. These spacecraft, dubbed ARTEMIS-P1 and ARTEMIS-P2, were tasked with investigating the Moon's interaction with solar wind, the acceleration of charged particles, and turbulence in the lunar space environment.

The changeover entailed a series of lunar flybys and deep-space excursions to prepare the spacecraft for lunar orbit insertion. ARTEMIS-P1 and ARTEMIS-P2 were carefully placed into stable orbits around the Moon's Lagrange points, affording a unique view of the interaction between the solar wind and the Moon.

The ARTEMIS mission built on the THEMIS project's legacy of improving space research. It sought to solve new issues concerning lunar space weather, such as how solar wind impacts the Moon and how the Moon's space environment effects the Earth's magnetosphere. By investigating these interactions, ARTEMIS hoped to improve our knowledge of space weather occurrences and their effects on Earth and the moon.

The Impact and Legacy

The THEMIS-ARTEMIS missions mark an important milestone in space exploration and our knowledge of the space environment. These missions' data have not only increased our understanding of the

Earth's magnetosphere and lunar space weather, but have also offered vital insights into the overall dynamics of space weather.

As we reflect on THEMIS and ARTEMIS' accomplishments, we acknowledge their contributions to space research and technology. These missions opened the stage for future exploration and laid the groundwork for comprehending the complicated interplay of solar winds, magnetic fields, and planetary bodies.

To summarize, the THEMIS-ARTEMIS project owes its existence to human curiosity and inventiveness. From investigating Earth's magnetosphere to pioneering investigation of the Moon's space environment, the mission has expanded our knowledge of space weather and its consequences on our planet. THEMIS and ARTEMIS left a legacy that continues to inspire future generations of scientists and explorers, reminding us of the universe's limitless potential.

As you read more into "ARTEMIS Unveiled: The Moon's Electric Shield," you'll learn about the incredible journeys of these missions and the findings that have transformed our knowledge of space. Join us as we uncover the mysteries of lunar space weather and the enormous influence of the THEMIS-ARTEMIS mission on our quest to comprehend the universe.

CHAPTER 2

The THEMIS Satellites

Launch and Deployment

The launch of the THEMIS satellites was more than just a countdown to liftoff; it was a painstakingly choreographed event that marked a new chapter in our effort to comprehend Earth's magnetosphere. As the five THEMIS satellites prepared for their trip, the world waited in anticipation, knowing that this mission would reveal new insights into the dynamics of space weather.

Launch Vehicle and Site

On February 17, 2007, the THEMIS mission launched from the historic SLC-17B launch site at Cape Canaveral, Florida. The five satellites were sent into orbit using a Delta II launch vehicle, which is noted for its dependability and track record. This vehicle, a workhorse of the space launch business, was selected for its ability to transport several payloads to orbit precisely.

The selection of Cape Canaveral's Space Launch Complex 17B was both symbolic and practical. This location, steeped in space exploration history, offered an outstanding vantage point for launching the THEMIS satellites on a trajectory that would allow them to properly investigate the Earth's magnetosphere. As the countdown hit zero and the rocket roared to life, the five THEMIS satellites set forth on their mission, with the prospect of discovering new cosmic mysteries.

Launch Delays and Rescheduling

However, the journey to space was not without hurdles. The THEMIS launch was originally slated for October 19, 2006, but was delayed many times. The principal difficulty resulted from workmanship issues with the Delta II's second stage, which had damaged the previous STEREO mission. As a result, the launch date was rescheduled until February 15, 2007.

However, the route to launch was beset with more delays. On February 13, 2007, poor weather caused a delay in fuelling the second stage. This mishap delayed the launch by 24 hours. On February 16, 2007, only minutes before the planned launch, a last weather balloon revealed adverse circumstances, prompting a pause at T-4 minutes. The crew started a 24-hour turnaround operation, aiming for a fresh launch window between 23:01 and 23:17 UTC on February 17.

Despite these challenges, the launch on February 17 went flawlessly. The Delta II rocket blasted off at 23:01 UTC, and the spacecraft separated from the launch vehicle after around 73 minutes. On February 18, mission operators at the Space Sciences Laboratory (SSL) at the University of California, Berkeley, verified successful contact with all five satellites, marking a victorious start to the THEMIS mission.

Satellite Design and Construction.

The THEMIS satellites were wonders of contemporary engineering, meticulously engineered to resist the harsh environment of orbit and conduct sophisticated scientific experiments. Their creation required a joint effort that brought together experts from many organizations.

The Function of Swales Aerospace

Swales Aerospace, a corporation known for its space system expertise, oversaw the building of the THEMIS satellites.

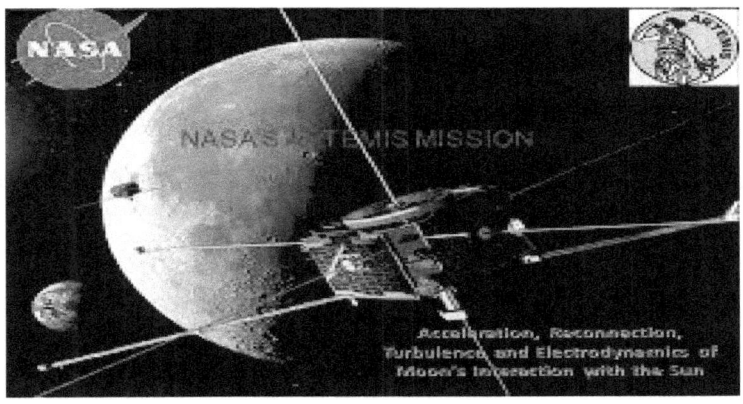

Swales Aerospace, which is based in Beltsville, Maryland, was instrumental in developing and producing the mission's five probes. Swales achieved a key milestone with this project, its second large satellite mission after the Earth Observing-1 (EO-1) spacecraft.

Swales Aerospace was in charge of integrating the satellites' essential components, which included the Battery Assembly Unit (BAU), Inertial Reference Unit (IRU), solar arrays, antennae and batteries. Each satellite was methodically manufactured and tested at the Beltsville plant before being sent to the University of California, Berkeley, for final integration with scientific equipment.

Instrumentation Overview

Each THEMIS satellite was outfitted with an array of advanced equipment intended to capture the intricacies of space weather and magnetospheric processes. These equipment were vital to the mission's success, providing the data required to answer important scientific questions.

- The Fluxgate Magnetometer (FGM) monitored the intensity and direction of magnetic fields in space. The FGM contributed to scientists' understanding of the structure and changes in the Earth's magnetosphere by collecting comprehensive magnetic field data.

- **Electrostatic Analyzer (ESA):** The ESA is intended to quantify the energy and density of charged particles. This data was critical for investigating the interactions between solar wind and the Earth's magnetosphere, since it provided information on particle acceleration and energy distribution.

- Solid State Telescope (SST): The SST detected high-energy particles in the space environment. Its findings were critical to understanding the behavior of these particles during geomagnetic substorms and magnetic reconnection occurrences.

- Search-Coil Magnetometer (SCM): The SCM detected magnetic field oscillations, which provided information about plasma waves and other dynamic processes in the magnetosphere. This data was critical for understanding temporal fluctuations in magnetic activity.

- **Electric Field Instrument (EFI):** The EFI monitored electric fields in space and provided data on how they interact with magnetic fields and charged particles. Scientists used this gadget to explore the magnetosphere's electrodynamics.

Testing & Integration

Before embarking on their mission, the THEMIS satellites underwent extensive testing to guarantee their suitability for orbit. This phase was essential in ensuring that the spacecraft performed as predicted in the harsh environment of space.

Thermal vacuum, vibration, and acoustic tests

The satellites were put through a series of tests meant to imitate the severe circumstances they would face in orbit. Thermal-vacuum tests evaluated the spacecraft's capacity to survive temperature changes and vacuum in space. Vibration testing guaranteed that the satellites could withstand the pressures of launch, while acoustic tests assessed their resistance to the high noise levels during liftoff.

These experiments were carried out at NASA's Jet Propulsion Laboratory (JPL) in Pasadena, California, a facility known for its competence in spacecraft testing. Swales Aerospace and JPL collaborated to guarantee that the THEMIS satellites were adequately equipped to face space difficulties.

Collaboration with NASA's Jet Propulsion Laboratory.

The partnership with NASA's Jet Propulsion Laboratory was critical to the THEMIS mission's success. JPL performed an important role in the satellites' testing and integration, employing its substantial knowledge and equipment to certify the spacecraft's performance.

This relationship went beyond testing to include many facets of mission planning and execution. JPL's experience in spacecraft operations and mission management helped the THEMIS satellites launch smoothly and begin their scientific mission successfully.

A Journey of Discovery

The launch and deployment of the THEMIS satellites signaled the start of an incredible voyage into space. Each satellite, precisely developed and tested, set off on a quest to discover the secrets of Earth's magnetosphere and beyond. The data obtained by these satellites would not only help us comprehend space weather, but would also pave the road for future exploration.

As you read further into "ARTEMIS Unveiled: The Moon's Electric Shield," you'll learn about the THEMIS mission's astonishing accomplishments and the critical role of its satellites in altering our knowledge of space. THEMIS' narrative, from launch obstacles to the complexities of satellite design and testing, is a monument to human creativity and the never-ending search of knowledge.

Join us as we continue to explore the intriguing realm of lunar space weather and the momentous discoveries that await us.

CHAPTER 3

The Mission Phases of THEMIS

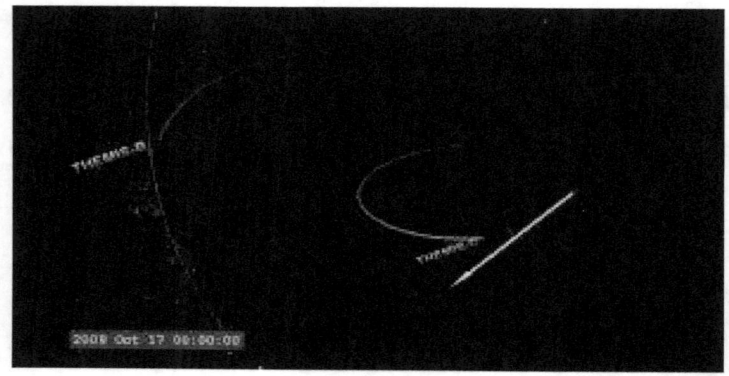

Initial Mission Phase

As the THEMIS satellites set off on their cosmic trip, the mission began with a meticulously organized dance of scientific investigation and orbital mechanics. This stage, called as the Initial Mission Phase, lay the framework for the mission's succeeding stages, allowing for a more in-depth research of the Earth's magnetosphere.

String of Pearls Configuration

The first mission phase was distinguished by the "String-of-Pearls" orbital arrangement. In this configuration, the five THEMIS satellites were placed in a line, deliberately spaced to form a string of pearls in orbit above Earth. This arrangement was critical to the mission's early scientific goals since it enabled the satellites to monitor and gather data from several sites along a single flight.

The String-of-Pearls arrangement was more than simply a technical option; it was also a planned tactic to maximize magnetosphere observation coverage. Each satellite in this configuration served as a distinct vantage point, offering a complete picture of the magnetospheric environment. Scientists might learn about the geographical and temporal fluctuations in magnetic fields and plasma dynamics by examining data from many satellites along this line.

During this phase, the major goal was to calibrate the equipment and fine-tune the observational parameters. The data acquired from this setup offered a baseline knowledge of the magnetosphere's activity and assisted in identifying significant areas of interest for future research. It was a critical step in developing the mission's scientific foundation and preparation for the more comprehensive observations that would come later.

The Dawn phase

After completing the Initial Mission Phase, the THEMIS mission entered the Dawn Phase. This stage saw a substantial change in the mission's emphasis, with the satellites transitioning to new orbits to allow for more extensive research of the Earth's magnetosphere.

Orbital Adjustments and Objectives

The Dawn Phase was distinguished by a purposeful modification of the satellite orbits. From September 15 to December 4, 2007, the THEMIS satellites were moved into more distant orbits, putting them on the dawn side of the magnetosphere. This orbital change enabled the spacecraft to examine the magnetosphere from a unique vantage point, since their orbits were positioned at apogee, the furthest point from Earth, on the magnetosphere's dawn side.

The goal of this phase was to provide the groundwork for detailed scientific observations and data gathering. The new orbits afforded an ideal view for investigating the interplay between solar wind and the Earth's magnetosphere. The dawn-side orientation enabled the satellites to study the formation and dynamics of charged particles, offering vital insights into the mechanisms that drive geomagnetic activity.

As the satellites settled into their new orbits, scientists concentrated on studying the magnetosphere's structure and behavior. The Dawn Phase helped to refine the mission's scientific aims and prepare for the Tail Science Phase, which would be more focused on exploration.

Tail Science Phase

As the THEMIS mission advanced, it reached the Tail Science Phase, which is critical for researching the Earth's magnetotail. This phase represented a transition from early observations to a more focused investigation of the magnetosphere's tail area, which was believed to include several significant phenomena of interest.

Magnetotail Exploration and Data Collection

The Tail Science Phase started on December 4, 2007, and was focused on investigating the magnetotail—an extended area stretching away from Earth where the solar wind interacts with the Earth's magnetic field. During this time, the THEMIS satellites' orbits were positioned at apogee inside the magnetotail, affording a unique view of this vital area.

The Tail Science Phase aimed to examine the complicated interactions between the solar wind and the magnetosphere's tail. This area is noted for its dynamic and turbulent activity, which includes the accumulation of magnetic energy and the occurrence of reconnection events. Scientists hoped that by locating the satellites in this area, they would be able to collect precise data on the magnetotail structure and behavior.

The data gathered during this phase were critical in understanding how energy is transmitted from the solar wind to the Earth's magnetosphere. It also gave information on the mechanisms that drive geomagnetic activity and the development of phenomena such as auroras.

Observations on Substorms and Magnetic Reconnection

One of the most interesting parts of the Tail Science Phase was the ability to closely monitor substorms and magnetic reconnection occurrences. Substorms are severe disruptions in the Earth's magnetosphere that cause auroras to brighten significantly. These occurrences are caused by the release of stored magnetic energy and may have serious consequences for space weather.

The THEMIS satellites gave scientists a front-row seat to these occurrences, enabling them to examine their evolution and effect. Researchers were able to study the beginning and growth of substorms by examining data from various satellites, providing insight into the fundamental processes driving these phenomena.

Magnetic reconnection, which occurs when magnetic field lines from the Sun and Earth reunite, was another focus of study. Reconnection events are responsible for the transport of energy and charged particles from the solar wind to the magnetosphere. The THEMIS mission's observations gave vital insights into how these events unfolded and how they influenced space weather.

The Tail Science Phase was a watershed moment in space research, providing a plethora of information that advanced our knowledge of the Earth's magnetosphere and its interactions with the solar wind. The findings from this phase have had a significant influence on our understanding of space weather and its implications on Earth.

The Legacy of THEMIS

The THEMIS mission's three phases—Initial, Dawn, and Tail

Science—represent a thorough investigation of the Earth's magnetosphere. Each step was methodically planned to answer particular scientific questions and get a better knowledge of space weather events.

As you read "ARTEMIS Unveiled: The Moon's Electric Shield," you'll learn about how the THEMIS mission's phases helped us understand space weather and geomagnetic dynamics. From strategic satellite placement to pioneering discoveries of substorms and magnetic reconnection, the THEMIS program exemplifies human curiosity and scientific inventiveness.

Join us as we continue to investigate the incredible journey of the THEMIS mission and its implications for our knowledge of the universe. The lessons learned throughout these mission stages have set the path for future discoveries and continue to inspire new generations of scientists and explorers.

CHAPTER 4

Groundbreaking Discoveries and Data

Magnetic Ropes and Solar-Terrestrial Interaction

In the vast, electric expanse of space, the Moon is more than just a silent observer. Its interaction with the Earth's magnetosphere, charged particles from the Sun, and the dance of magnetic fields tell a fascinating story that we are only now beginning to unravel. This narrative revolves around the concept of magnetic ropes, also known as flux ropes. These fascinating structures are essential in the complex dance between the Sun's charged particles and the Earth's magnetic shield.

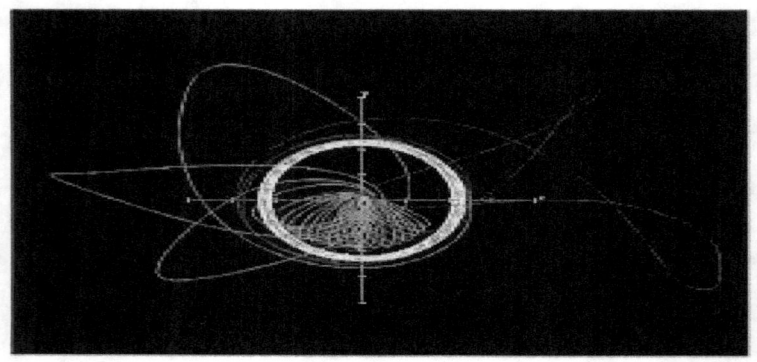

Evidence for Birkeland Currents

The existence of Birkeland currents, named after pioneering Norwegian scientist Kristian Birkeland, sheds light on the magnetic connection between the Sun and the Earth. These currents, which consist of streams of charged particles, flow along magnetic field lines and are a critical component of the solar-terrestrial interaction. Birkeland's pioneering experiments in the early twentieth century laid the groundwork for our current understanding of how charged particles travel through space, influencing everything from auroras to electrical grids on Earth.

Moving forward to the modern era, we have sophisticated missions like ARTEMIS that are designed to study these phenomena in unprecedented depth. ARTEMIS data confirmed the presence of Birkeland currents that extend to the Moon. This discovery has far-reaching implications, implying that the Moon, too, plays a role in the intricate dance of solar and terrestrial magnetism. It has been discovered that these currents can span vast distances, connecting the Earth's magnetosphere to the Moon's environment.

NASA's "30-kilovolt battery" comparison

To make the concept of these magnetic interactions more tangible, NASA scientists have frequently compared the process to a "30 kilovolt battery" in space. This metaphor depicts the enormous energy transferred by these currents as they travel through the Earth's magnetosphere and beyond. Imagine a massive battery in the sky that is charged and discharged by solar winds and magnetic reconnections. This analogy helps to convey the scope and intensity of the phenomena that scientists are investigating.

The ARTEMIS mission has provided critical data supporting this comparison, demonstrating how energy is stored and released in the magnetosphere, similar to a battery cycle. As the solar wind moves past Earth, it carries magnetic field lines with it, stretching them until they snap back, releasing stored energy in a burst. This snapping and reconnecting of magnetic field lines is not a theoretical concept, but rather a dynamic and observable process with tangible consequences.

Triggering of Magnetospheric Substorms

The magnetospheric substorm is one of the most dramatic examples of solar-terrestrial interaction. These events, which are distinguished by abrupt and intense changes in the magnetosphere, can result in spectacular auroras and significant disruptions to satellites and communication systems. Understanding what causes these substorms is critical, not only for scientific reasons, but also for the practical implications of predicting and mitigating their effects.

The Role of Magnetic Reconnection

Magnetic reconnection is the driving force behind these substorms. This phenomenon occurs when opposing magnetic field lines come into contact and rearrange, releasing massive amounts of energy. Consider two stretched rubber bands snapping and rejoining; the release of tension is analogous to the energy released during magnetic reconnection.

The ARTEMIS mission has been instrumental in observing these reconnections near the Moon, providing new information about how and why they occur. Before ARTEMIS, many people assumed that reconnection was primarily an Earth-bound phenomenon. However, evidence gathered from lunar orbit has broadened our understanding, revealing that reconnection events can occur further away, affecting a larger area than previously believed.

Case Study: February 26, 2008 Discovery.

One of the most significant breakthroughs occurred on February 26, 2008. On this date, ARTEMIS detected an unexpected and powerful reconnection on the Moon's far side. This discovery was a watershed moment, calling into question existing models and assumptions about the spatial extent of magnetospheric activity.

The data collected revealed an intense magnetic rope formation connecting the Earth's magnetosphere to the Moon, confirming that magnetic reconnection was taking place at a much greater distance from Earth than expected. The implications were staggering: it was now clear that the Moon could influence the dynamics of Earth's space weather system.

The February 2008 discovery allowed scientists to reconsider the limits of the Earth's magnetosphere and the Moon's influence. It demonstrated how the Moon serves as a conduit for solar-terrestrial energy, actively participating in the interplay of magnetic forces that shape our planet's space weather.

The Moon's Role in Space Weather.

The ARTEMIS mission, which investigated the Moon's role in space weather, revealed a dynamic and complex system of interactions. The discovery of magnetic ropes, evidence of Birkeland currents, and the demonstration of magnetic reconnection near the Moon are all changing our perception of the solar-terrestrial relationship.

These ground-breaking discoveries highlight the Moon's active participation in space weather phenomena, challenging the notion that our lunar neighbor is a passive bystander. As our understanding grows, so does our appreciation for the intricate and beautiful dance of cosmic forces that shape our planet.

With each new piece of information, we get closer to fully comprehending the Moon's role in the solar system's grand story. The ARTEMIS mission continues to shed light on these hidden dynamics, revealing the Moon's secrets and expanding our understanding of the universe.

As we delve deeper into this fascinating field, we discover that the Moon is more than just a distant companion; it is an essential player in the cosmic orchestra of magnetic forces. The more we learn, the more we realize we are part of a much larger, interconnected system, with the Moon playing a role in the universal symphony of cosmic weather.

CHAPTER 5

Transition to ARTEMIS

The ARTEMIS mission marks a watershed moment in our investigation of the Moon and its surroundings. ARTEMIS (Acceleration, Reconnection, Turbulence, and Electrodynamics of the Moon's Interaction with the Sun) was born from a visionary repurposing of two existing spacecraft. It transitioned from its original mission of observing the Earth's magnetosphere as part of NASA's THEMIS mission to a pioneering lunar science expedition. This chapter looks into the mission's goals, the technical obstacles of transferring the spacecraft, and the complex maneuvers that placed ARTEMIS in orbit around the Moon.

ARTEMIS Mission Objectives

ARTEMIS was designed with a lofty goal in mind: to unveil the mysteries of lunar space weather and to better comprehend the Moon's function in the larger framework of the Sun-Earth connection. As our understanding of the Moon's magnetosphere grew, the necessity for a specialized mission became clear. ARTEMIS aims to examine three main scientific objectives:

1. Exploring Lunar-Solar Interactions: ARTEMIS aims to better understand how the Moon interacts with the solar wind, a stream of charged particles generated by the Sun. These interactions may result in intricate magnetic structures surrounding the Moon, impacting its surface and environment.

2. Plasma Dynamics: The mission's goal is to explore plasma dynamics in the lunar environment, particularly the behavior of charged particles and magnetic fields. Understanding these processes is critical to appreciating the Moon's impact on space weather.

3. Exploring the Lunar Exosphere and Magnetotail: ARTEMIS is charged with investigating the Moon's thin atmosphere, or exosphere, and its function in the Earth's magnetotail, which is the area where the solar wind draws the Earth's magnetic field away. This investigation may provide fresh insights on the Moon's influence on the Earth's magnetosphere.

By fulfilling these goals, ARTEMIS aims to increase our knowledge of the Moon's role in the solar system's electromagnetic symphony while also providing insights into the obstacles and potential for future lunar research.

Lunar and Earth Flybys

ARTEMIS' trip to the Moon was an incredible achievement of engineering and planning. The spacecraft was originally launched as part of the THEMIS mission, with the main purpose of researching Earth's auroras. However, as the project progressed, NASA scientists saw a chance to repurpose two of the five THEMIS spacecraft—ARTEMIS-P1 and ARTEMIS-P2—for lunar research.

Challenges of Transition

Repurposing the THEMIS spacecraft for the ARTEMIS mission presented major obstacles. The spacecraft were initially meant to operate in the Earth's magnetosphere, not in the harsh conditions of lunar orbit. To complete the mission, scientists and engineers had to invent novel techniques for conserving fuel, managing heat conditions, and ensuring reliable communication across huge distances to the Moon.

Lunar flybys

To save fuel and assure a seamless transition, ARTEMIS performed a series of lunar flybys. These procedures entailed leveraging the Moon's gravitational field to change the spacecraft's course, gradually accumulating orbital energy in preparation for insertion into lunar orbit. The flybys required careful calculations and coordination to guarantee that ARTEMIS approached the Moon at the proper angle and speed.

During these flybys, ARTEMIS collected crucial data on the lunar environment, providing early insights of the mission's prospective findings. These earliest contacts revealed details about the Moon's magnetic field, surface composition, and plasma environment, laying the groundwork for the mission's core scientific goals.

Earth fly-bys

In addition to lunar flybys, ARTEMIS used Earth flybys to change its route and save fuel. These maneuvers enabled the spacecraft to collect energy from Earth's gravity and change their orbits without using precious fuel. The combination of Earth and lunar flybys demonstrated the mission team's skill and resourcefulness in navigating difficult gravitational settings with accuracy and efficiency.

Insertion into lunar orbit

After months of painstaking preparation and execution, ARTEMIS accomplished a major milestone: insertion into lunar orbit. This phase of the mission represented a fresh beginning, as the spacecraft switched from their dynamic trip through space to their main scientific purpose.

ARTEMIS-P1's orbital maneuvers

ARTEMIS-P1 was the first of the two spacecraft to enter lunar orbit. This procedure included a series of sophisticated maneuvers to alter the spacecraft's trajectory and velocity, guaranteeing a stable and efficient orbit around the moon. During this phase, the mission crew encountered various problems, including the need to carefully manage fuel use while still maintaining connection with Earth.

ARTEMIS-P1 had to perform a series of precisely scheduled engine fires throughout the insertion phase in order to progressively reduce its velocity and enable the Moon's gravity to catch the spacecraft. These movements were executed with precision, using the mission team's skills and experience to guarantee a smooth transition.

Once in orbit, ARTEMIS-P1 commenced scientific measurements of the Moon's magnetic environment and plasma dynamics. The spacecraft's equipment offered previously unseen insights into the interactions between the Moon and the solar wind, giving new information on the Moon's role in the solar system's electromagnetic dance.

ARTEMIS-P2's Mission to Lunar Orbit

Following the successful insertion of ARTEMIS-P1, emphasis shifted to ARTEMIS-P2's mission to lunar orbit. This spacecraft encountered unusual problems that necessitated a special series of maneuvers to attain its orbital goals.

ARTEMIS-P2's voyage entailed a series of intricate gravitational interactions, with both the Moon and the Earth helping to change its course. These movements need perfect timing and synchronization, since even minor deviations may jeopardize the spacecraft's ability to enter lunar orbit.

Despite the hurdles, ARTEMIS-P2 successfully reached lunar orbit, joining its counterpart in the mission's scientific pursuits. Together, the two spacecraft constituted a formidable observational platform capable of examining the Moon's environment from numerous perspectives and delivering detailed data on its magnetic and plasma processes.

Achieving Lagrange Points Orbit

One of the most ambitious objectives of the ARTEMIS mission was to enter orbit around the Earth-Moon Lagrange points. These are spots in space where the gravitational influences of the Earth and Moon are balanced, resulting in stable zones where spacecraft may hold position with little fuel usage.

Significance of Lagrange Points

The Earth-Moon Lagrange points provide unique possibilities for scientific discovery and observation. These locations, which are strategically located between the Earth and the Moon, give an excellent vantage point for examining the interactions between the two celestial bodies and their environs.

For ARTEMIS, entering orbit around the Lagrange points was an important milestone since it allowed the spacecraft to perform long-term measurements of the Moon's magnetosphere and its interactions with the solar wind. This capacity enabled the mission to gather data over prolonged time periods, offering vital insights into the dynamic processes that shape the lunar environment.

Navigating to Lagrange points

Reaching the Lagrange points required a series of precise movements as the spacecraft balanced gravity pulls and maintained a stable orbit. These moves required meticulous preparation and execution, with mission scientists relying on their knowledge to traverse the challenging gravitational terrain.

Once near the Lagrange points, ARTEMIS made a number of important findings that shed insight on the Moon's magnetic interactions and plasma dynamics. The data gathered during this part of the mission improved our knowledge of the Moon's involvement in the larger context of space weather and its effect on the Earth's magnetosphere.

Conclusion: The Artemis Legacy

The ARTEMIS project is a spectacular success in space research, converting two Earth-orbiting spacecraft into lunar scientific pioneers. ARTEMIS overcame major hurdles to fulfill its scientific goals, revealing new information about the Moon's magnetic environment and interactions with the solar wind.

As we reflect on the mission's successes, it is apparent that ARTEMIS has had a lasting impact on our knowledge of the Moon and its role in the solar system's electromagnetic dynamics. The mission's legacy continues to inspire future generations of scientists and engineers, inspiring them to investigate the wonders of space and discover our cosmic neighbors.

With each new discovery, ARTEMIS broadens our understanding of the Moon and its role in the universe, reminding us of the limitless potential of exploration and the timeless appeal of the cosmos. As we look forward, the lessons learnt from ARTEMIS will influence our attempts to travel farther into the unknown, paving the way for new experiences and discoveries beyond the horizon.

CHAPTER 6

The ARTEMIS Lunar Operations

The ARTEMIS mission is a huge step forward in our effort to comprehend the Moon's role in the cosmic dance of the Sun, Earth, and space weather. With ARTEMIS-P1 and ARTEMIS-P2 safely entered lunar orbit, the mission starts its operational phase, ready to discover fresh insights into the Moon's ecology and its interaction with solar events. This chapter delves into ARTEMIS' complex operations in lunar orbit, the mission's primary scientific aims and accomplishments, and the intriguing possibilities for future lunar exploration.

ARTEMIS-P1 and ARTEMIS-P2 are in lunar orbit.

The successful deployment of ARTEMIS-P1 and ARTEMIS-P2 into lunar orbit signaled the start of an unparalleled investigation of the Moon's magnetic field. Each satellite has a distinct orbital route, allowing them to conduct complementary studies and offer a complete picture of lunar plasma dynamics.

ARTEMIS-P1: A New Perspective.

ARTEMIS-P1 entered a very elliptical orbit around the Moon, allowing it to explore various areas of the lunar surface. Its orbit is intended to provide maximum coverage of the Moon's magnetosphere, allowing ARTEMIS-P1 to investigate how solar wind interacts with the Moon at different distances and angles. This strategic orientation provides ARTEMIS-P1 with a unique vantage point for studying the Moon's magnetic tail and its interactions with the Earth's magnetosphere.

ARTEMIS-P1 has collected useful data on the behavior of charged particles, magnetic field structures, and plasma waves throughout its orbit. These findings help scientists understand how the Moon functions as a shield, diverting solar wind and shielding its surface from hazardous radiation.

ARTEMIS-P2: Exploring Lunar Mysteries.

While ARTEMIS-P1 studies the Moon's magnetosphere, ARTEMIS-P2 investigates the lunar surface and its interactions with space. ARTEMIS-P2, which has a slightly different orbit, gives important information on the Moon's exosphere and surface composition.

ARTEMIS-P2's sensors are intended to examine the composition and density of the Moon's thin atmosphere, as well as the interactions between solar wind and the lunar surface. These observations provided fresh information regarding the Moon's surface features, such as the existence of volatile substances and the impacts of space weathering.

Key Scientific Goals and Achievements

The ARTEMIS mission aimed to address basic concerns concerning the Moon's magnetic environment and its function in the solar-terrestrial system. As the mission proceeds, ARTEMIS has made substantial progress toward its scientific objectives, giving important insights into the mechanics of lunar space weather.

Understanding Lunar and Solar Interactions

One of the main goals of the ARTEMIS mission is to investigate how the Moon interacts with the solar wind. The Moon lacks a global magnetic field like Earth, therefore its surface is directly exposed to the stream of charged particles from the Sun. ARTEMIS has been essential in analyzing these interactions, showing how the Moon's localized magnetic fields, known as magnetic anomalies, influence the behavior of solar wind particles.

ARTEMIS investigations have shown that these magnetic anomalies form mini-magnetospheres on the lunar surface, which may deflect solar wind and generate areas of increased or decreased particle activity. This finding has important implications for comprehending the Moon's surface environment and the possibility of protecting future lunar homes from hazardous radiation.

Investigating Plasma Dynamics

Plasma dynamics play an important role in the lunar environment, impacting everything from magnetic field structures to surface charge. ARTEMIS has produced unparalleled data on plasma wave behavior, magnetic reconnection events, and charged particle transit near the Moon.

One of ARTEMIS' significant discoveries is the discovery of plasma waves caused by the interaction between solar wind with the Moon's surface. These waves may travel over the lunar environment, carrying energy and momentum that can affect the Moon's surface and exosphere. Understanding these plasma dynamics is critical for anticipating the Moon's reaction to solar activity and the implications for future lunar missions.

Exploring the Moon's Exosphere and Magnetotail

The Moon's exosphere is a thin atmosphere made up of sparse atoms and molecules that are impacted by solar radiation and particle bombardment. ARTEMIS has supplied important data on the composition and dynamics of this exosphere, giving new information about its behavior and development.

ARTEMIS also investigated the Moon's function in the Earth's magnetotail, which is the area where the solar wind extends the Earth's magnetic field. The mission examined how the Moon interacts with this area, giving information on the passage of energy and particles between Earth and Moon. These observations have helped us better comprehend the Moon's effect on the Earth's space environment, as well as its potential impact on space weather.

Key Discoveries and Scientific Achievements

Since the start of its lunar activities, the ARTEMIS mission has made several scientific advances that have altered our knowledge of the Moon and its environment.

- Magnetic Anomalies and Mini-Magnetospheres: ARTEMIS has mapped the Moon's magnetic anomalies, showing how they form localized mini-magnetospheres that deflect solar wind and affect surface processes.

- Plasma Waves and Magnetic Reconnection Events: The mission discovered plasma waves and magnetic reconnection events near the Moon, revealing new information on the dynamics of lunar space weather and its influence on the Moon's environment.

- Lunar Exosphere Composition: ARTEMIS assessed the composition and density of the Moon's exosphere, providing new information on its behavior and development in response to solar activity.

- Interactions with the Earth's Magnetotail: The mission investigated the Moon's interactions with the Earth's magnetotail, showing how energy and particles are exchanged between the Earth and Moon and how they affect space weather.

Long-term Mission Prospects

As the ARTEMIS mission continues to operate, the possibilities for future lunar discoveries and breakthroughs seem excellent. The mission's data and conclusions are influencing the future of lunar exploration, opening the way for subsequent missions and projects to better understand the Moon's ecosystem.

Increasing our understanding of lunar space weather.

ARTEMIS has provided the basis for a better understanding of lunar space weather and what it means for future lunar missions. The mission's data has offered vital insights into the behavior of solar wind, plasma dynamics, and magnetic interactions in the lunar environment, which will help shape future missions and technology.

As scientists evaluate the ARTEMIS data, new models and ideas are emerging to forecast and alleviate the consequences of space weather on the Moon. These developments are important to the safety and success of future lunar missions, including human exploration and habitation.

Inspiring New Lunar Missions

The success of ARTEMIS has sparked a new surge of interest in lunar exploration, with agencies and groups throughout the globe planning missions to investigate the Moon's environment and resources. ARTEMIS' discoveries have underlined the necessity of understanding the Moon's magnetic and plasma dynamics, which will shape the design and aims of future forthcoming missions.

Future missions will expand on ARTEMIS' findings by investigating other areas of the Moon and undertaking experiments to validate hypotheses and models built using ARTEMIS data. These missions will help us better understand the Moon's environment and its ability to support human activities.

The Promise of Lunar Exploration

The ARTEMIS mission has illustrated how lunar exploration may answer basic concerns regarding the Moon and its place in the solar system. As we look forward, the lessons acquired from ARTEMIS will drive our attempts to study the Moon and uncover its mysteries.

Lunar exploration presents the possibility of fresh scientific discoveries and technological improvements, with the ability to revolutionize our knowledge of the solar system and our role within it. The ARTEMIS mission has set the stage for a new era of exploration, encouraging future generations to aim for the Moon and beyond.

Conclusion: The Legacy of Artemis

The ARTEMIS mission has had exceptional success in exploring the Moon's environment, giving significant data and insights that have helped us better comprehend lunar space weather. Through its innovative operations and scientific breakthroughs, ARTEMIS has created the groundwork for future missions and projects to investigate the Moon and its secrets.

As we reflect on ARTEMIS' legacy, we are reminded of the potential of exploration and discovery to broaden our understanding and inspire new possibilities. The mission's results continue to alter our knowledge of the Moon and its role in the solar system, directing our efforts to investigate and exploit the potential of our celestial neighbor.

With each new finding, ARTEMIS adds to our knowledge of the Moon and its role in the universe, reminding us of the perennial attraction of exploration and the limitless possibilities of space. As we look forward, the lessons acquired from ARTEMIS will guide our voyage to the Moon and beyond, paving the way for new experiences and discoveries in the great expanse of space.

CHAPTER 7

Collaborative Efforts and International Contributions

The ARTEMIS mission exemplifies the potential of teamwork, bringing together the knowledge and resources of several countries and organizations to investigate the Moon's electric shield. This chapter digs into the multinational contributions and collaborations that have helped make ARTEMIS a reality, emphasizing the crucial roles of Canada, Austria, Germany, and France, as well as NASA's Launch Services Program and the University of California,

Berkeley. Together, these partnerships have driven ARTEMIS to unprecedented heights of lunar research.

Contributions by Canada, Austria, Germany, and France

The ARTEMIS mission has greatly benefitted from the knowledge and technical breakthroughs of its worldwide partners. Each nation has contributed distinct assets and skills to the project, helping it succeed and further our knowledge of lunar space weather.

Canada's expertise in space instrumentation.

Canada has long been a pioneer in space equipment, providing superior scientific instruments for space research. Canadian scientists and engineers contributed significantly to the ARTEMIS program by developing instruments and data processing tools.

One of the most significant Canadian contributions was the creation of high-precision magnetometers, which are critical for detecting the Moon's magnetic field and plasma environment. These sensors have allowed ARTEMIS to gather extensive data on the Moon's magnetic anomalies and their interactions with the solar wind, revealing new information about the dynamics of lunar space weather.

Canadian scientists have also contributed significantly to the interpretation of ARTEMIS data and the development of models to better understand the Moon's magnetic interactions. This partnership has increased the mission's scientific output and our understanding of the lunar environment.

Austria's Contribution to Plasma Physics

Austria has a strong research record in plasma physics, which is fundamental to the ARTEMIS mission's aims. Austrian scientists helped build plasma analyzers and sensors that detect the behavior of charged particles near the Moon.

These experiments have supplied crucial information on plasma dynamics in the lunar environment, showing how solar wind interacts with the Moon's surface and magnetic field. Austrian academics have played a key role in evaluating this data, resulting in fresh insights on the Moon's involvement in the larger solar-terrestrial system.

The Austrian team's plasma modeling skills has also helped to produce models that forecast the behavior of plasma waves and magnetic reconnection occurrences near the Moon. These models helped shape the mission's scientific goals and directed its investigation of lunar space weather.

German Innovation in Spacecraft Technology

Germany has a long history of innovation in spacecraft technology, and its contributions to the ARTEMIS mission were vital to its success. German engineers and scientists contributed expertise in spacecraft architecture, propulsion systems, and communication technologies, allowing ARTEMIS to negotiate the difficult gravitational environment of the Earth-Moon system.

One of Germany's most significant contributions was the development of sophisticated propulsion systems, which allowed ARTEMIS to undertake the exact maneuvers necessary for lunar orbit insertion and Lagrange point operations. These mechanisms have enabled ARTEMIS to meet its orbital goals with exceptional efficiency, maximizing its scientific yield.

Germany also helped to develop communication systems, ensuring that ARTEMIS could maintain consistent contact with mission control and send crucial data back to Earth. This feature has proved critical to the mission's success, allowing continuous monitoring and study of the lunar environment.

France's Experience in Data Processing and Analysis

France has a long history of skill in data processing and analysis, and its contributions to the ARTEMIS project have helped translate raw data into relevant scientific discoveries. French scientists contributed significantly to the development of algorithms and software tools for processing ARTEMIS data, allowing the mission to learn more about the Moon's magnetic and plasma environments.

One of the most significant French contributions was the creation of data processing pipelines intended to manage the massive amounts of data produced by ARTEMIS devices. These pipelines have allowed real-time lunar analysis, yielding crucial insights into the dynamics of solar wind and magnetic fields.

French researchers have also worked on constructing models to understand ARTEMIS data, giving new information about the Moon's interactions with the solar wind and its involvement in space weather dynamics. These models helped shape the mission's scientific goals and guided its research of the Moon's electric shield.

The Role of NASA Launch Services Program (LSP)

The ARTEMIS mission would not have gone successful without NASA's Launch Services

Program (LSP). LSP was important in securing the safe and efficient launch of the ARTEMIS spacecraft, providing the technical knowledge and resources required to meet the mission's goals.

Ensure a Successful Launch.

The LSP team was in charge of supervising the launch of the ARTEMIS satellite, ensuring that the mission followed all technical and safety standards. This included negotiating with launch providers, overseeing the spacecraft's integration with the launch vehicle, and completing thorough testing to guarantee that the mission could endure the rigors of space flight.

The LSP team collaborated closely with the ARTEMIS mission team to create a thorough launch plan that included the actions necessary to safely and effectively deploy the spacecraft into its target orbit. This strategy comprised intensive simulations and rehearsals aimed at identifying and mitigating possible risks and ensuring a successful launch.

Supporting the Transition to Lunar Operations.

Following the successful launch of the ARTEMIS spacecraft, the LSP crew proceeded to assist with the transition to lunar operations. This involved communicating with mission control to monitor the spacecraft's performance, ensuring that it stayed on track and performed as intended.

The LSP team's knowledge in orbital dynamics and spacecraft operations was crucial during this phase of the project, advising on the maneuvers needed to accomplish lunar orbit insertion and Lagrange point operations. Their assistance meant that ARTEMIS made the successful transfer from its initial orbit to its operational configuration, allowing the mission to begin its scientific investigation of the Moon.

Integration and Support from the University of California, Berkeley

The University of California, Berkeley, has played a critical role in the ARTEMIS mission, providing the scientific leadership and technological skills required to realize the ambitious goals. Berkeley's Space Sciences Laboratory (SSL) has been at the vanguard of the mission's planning and operations, directing scientific research into the Moon's environment.

Scientific Leadership and Instrumentation

Berkeley's SSL has played a key role in designing the ARTEMIS mission's scientific goals, as well as building the equipment and technology needed to investigate the Moon's magnetic and plasma environments. Many of the mission's major equipment, like as magnetometers, plasma analyzers, and particle detectors, were conceived and constructed by the SSL team, allowing for thorough data collection on the lunar environment.

These devices have helped ARTEMIS fulfill its scientific objectives, offering unparalleled insights into the Moon's magnetic interactions and plasma dynamics. The SSL team's skill in equipment and data processing was vital to the mission's success, leading its investigation of lunar space weather and revealing new information about the Moon's role in the solar-terrestrial link.

Data Analysis and Interpretation

Berkeley's SSL has also been essential in processing and interpreting ARTEMIS data, converting raw measurements into useful scientific findings. The SSL team has developed sophisticated data processing algorithms that allow for real-time study of the lunar environment and disclose new information on the Moon's interactions with the solar wind.

The SSL team's data analysis experience has been useful in constructing models to better understand the behavior of magnetic fields and plasma waves in the lunar environment.

These models shaped the mission's scientific goals and led its research of the Moon's electric shield, revealing new information about the dynamics of lunar space weather.

Education and Outreach Initiatives

In addition to its scientific achievements, Berkeley's SSL has actively participated in educational and outreach activities, motivating the next generation of scientists and engineers to investigate the wonders of space. The SSL team has created educational programs and resources to help schools and the general public understand the science behind the ARTEMIS mission and its Moon exploration.

These activities have included public talks, seminars, and online materials, allowing students and educators to learn about the mission and its findings. The SSL team's dedication to education and outreach has increased awareness of the ARTEMIS mission and its contributions to lunar research, motivating a new generation of explorers to aspire for the sky.

Conclusion: The Power of Collaboration.

The ARTEMIS mission exemplifies the potential of teamwork, bringing together the knowledge and resources of several countries and organizations to investigate the Moon's electric shield. Canada, Austria, Germany, France, NASA's Launch Services

Program, and the University of California, Berkeley, have all made significant contributions to completing the mission's goals, which have advanced our knowledge of lunar space weather and its role in the solar-terrestrial link.

As we consider the combined efforts that have made ARTEMIS a reality, we are reminded of the value of collaborations in furthering our understanding of the universe. The mission's accomplishment highlights the importance of international cooperation in space research, showcasing the possibility of future partnerships unlocking fresh discoveries and inspiring the next generation of explorers.

With each new discovery, ARTEMIS adds to our knowledge of the Moon and its role in the universe, reminding us of the limitless potential of exploration and the timeless appeal of the cosmos. As we look to the future, the lessons acquired from ARTEMIS will drive our efforts to explore the Moon and beyond, paving the way for new experiences and discoveries in space.

CHAPTER 8

Future Prospects and Implications

As we look forward, the moon harbors undiscovered insights that promise to transform our knowledge of lunar space weather. The ARTEMIS mission has already offered tantalizing glimpses of these possibilities, providing a foundation for future research.

Hidden Dimensions of the Lunar Plasma Environment.

One of the most fascinating future research opportunities is a thorough examination of the three-dimensional nature of the lunar plasma environment. ARTEMIS has shown that this environment is dynamic and complicated, impacted by the moon's interactions with the solar wind and the Earth's magnetotail. Future missions might concentrate on recording plasma flows and electric fields surrounding the moon at great resolution, revealing previously unknown patterns and structures.

The discovery of novel plasma phenomena—perhaps similar to auroras on

Earth but unique to the lunar context—could transform our knowledge of how celestial entities without atmospheres interact with space weather. These findings would not only improve our understanding of the moon, but might also shed light on the behavior of other airless worlds in our solar system and beyond.

Unraveling the mysteries of lunar volatiles.

The finding of water ice and other volatiles in permanently shadowed areas of the moon has important implications for future lunar research and colonization. However, the interaction between these volatiles and the lunar space weather environment is mainly unknown. Future missions might investigate how solar wind and magnetotail interactions influence the stability and dispersion of these volatiles.

Understanding these processes may pave the road for sustainable lunar resource exploitation. Imagine future lunar outposts tapping into these water ice reserves to provide astronauts with crucial life support and fuel for greater space exploration. These improvements would not only make lunar living possible, but would also establish the moon as an important stepping stone for human exploration of the solar system.

Lunar Magnetism: A View Into Planetary Evolution

The moon's primordial magnetic field, which has mostly disappeared, contains clues to the solar system's past. Future missions might investigate the relics of this magnetic field in more detail, attempting to understand its genesis and death. Scientists can recreate the moon's magnetic history by examining lunar rocks and regolith, which provides insights into the processes that produced not just our moon but also other planetary bodies.

The ramifications of these findings go beyond scholarly interest. Understanding the moon's magnetic history may help us understand the evolution of Earth's magnetic field and its function in shielding our planet from damaging cosmic radiation—a factor important to the creation and maintenance of life.

The Impact of Lunar Space Weather Studies

The advances made possible by ARTEMIS have already paved the way for a new era in lunar space weather research. As we continue to build on these foundations, the implications for scientific inquiry and practical applications will be enormous.

Developing Predictive Models

One of the most notable achievements of ARTEMIS has been the development of forecast models for lunar space weather. These models, revised using data obtained by the mission, enable scientists to anticipate solar wind conditions and their impacts on the lunar surface with unparalleled precision. As future missions collect more data, these models will improve, allowing us to predict and reduce the impacts of space weather on lunar activities.

Accurate forecasts of space weather events will be critical for future lunar trips, safeguarding both the equipment and the explorers. The capacity to predict solar storms and radiation surges will allow lunar colonies to function securely and efficiently, reducing the hazards associated with extended exposure to cosmic and solar radiation.

Enhanced Communication and Navigation Systems

The discoveries provided from ARTEMIS have practical significance for lunar communication and navigation systems. Understanding the lunar space weather environment enables engineers to create more resilient communication networks, assuring dependable links between Earth and lunar outposts. Furthermore, updated simulations of the moon's plasma environment increase the precision of navigation systems, which is critical for both robotic and crewed missions.

These developments are not only for lunar uses. The methods created to overcome lunar space weather difficulties may be applied to other planetary bodies, expanding our potential to explore the solar system. From Mars to asteroids, the lessons acquired from ARTEMIS will guide future exploration endeavors, ensuring that humanity's reach extends far beyond Earth.

Lessons and Technological Advances

The ARTEMIS mission produced critical insights and technical advancements that will influence the future of space exploration. These findings are not only important for lunar exploration, but they also have larger implications for our knowledge of space weather and its effects on human activities.

Innovative Instrumentation and Data Analysis Techniques

ARTEMIS pioneered the use of cutting-edge technology to research the lunar environment, using advanced sensors and detectors to monitor electric and magnetic fields, plasma waves, and energetic particles. These devices have established new benchmarks for space weather research, emphasizing the significance of high-resolution, multi-point observations in comprehending complicated systems.

Future missions will expand on these breakthroughs, using even more sophisticated sensors to investigate the lunar environment. The development of smaller sensors and autonomous data processing methods will allow missions to collect and analyze massive volumes of data more quickly, hastening the rate of discovery.

Interdisciplinary Collaboration and Knowledge Sharing

The success of ARTEMIS demonstrates the value of multidisciplinary cooperation in space exploration. By bringing together professionals from several sectors, the expedition was able to tackle challenging scientific concerns and overcome technological obstacles. This collaborative approach will be critical for future missions, enabling the sharing of ideas and the creation of creative solutions.

The lessons from ARTEMIS underscore the need of open data and knowledge exchange. By making data available to academics throughout the globe, the mission has motivated a new generation of scientists and engineers to contribute to the area of space weather research. This democratization of information will spur innovation and guarantee that the advantages of space exploration are enjoyed by everyone.

Planning for Human Exploration

Perhaps the most important aspect of ARTEMIS is its role in preparing for human exploration of the moon and beyond. The mission has offered crucial insights into the problems and possibilities that come with living and working on the moon. From radiation protection to resource utilization, ARTEMIS has paved the way for future missions that will push the limits of human exploration.

As we look forward, the lessons acquired from ARTEMIS will inform the development of technologies and methods that will allow people to flourish on the moon. These advancements will not only make lunar exploration easier, but will also pave the way for humanity's trip to Mars and beyond.

Vision for the Future

Finally, the ARTEMIS mission has contributed to a better knowledge of lunar space weather and its implications for future exploration. The discoveries and breakthroughs enabled by ARTEMIS have paved the way for a new era of scientific inquiry and technological innovation.

As we continue to investigate the moon and other celestial bodies, the lessons acquired from ARTEMIS will guide our efforts, ensuring that we take full advantage of the potential provided by space exploration. By expanding on the groundwork set by this pioneering expedition, we may uncover the mysteries of the universe and extend humanity's reach into the cosmos.

The moon, long a faraway and enigmatic thing in the night sky, is now within reach. With each new discovery, we get closer to a future in which the moon serves as a portal to the stars, allowing mankind to live, work, and fantasize about what is beyond. As we begin on this voyage, the spirit of ARTEMIS will continue to inspire us, paving the path for a better future among the stars.

APPENDICES

The book's appendices provide more technical information, critical results, and vital resources that help us better appreciate the ARTEMIS mission and its breakthrough contributions to lunar space weather research. These appendices provide a thorough reference for those interested in learning more about the mission and the scientific findings that resulted from it.

Appendix A: Technical Specifications for THEI and ARTEMIS Satellites

The ARTEMIS mission, which evolved from the older THEMIS project, is an impressive accomplishment of engineering and scientific design. This appendix describes the technical specs of both the THEMIS and ARTEMIS satellites, emphasizing the technological advances that permitted the mission's success.

THEMIS Satellites.

The Time History of Events and Macroscale Interactions during Substorms (THEMIS) project was made up of five identical probes meant to investigate the Earth's magnetosphere and its interactions with the solar wind. Each satellite was outfitted with a set of equipment designed to study various elements of space weather occurrences.

1) Instruments:

- Electric Field Instrument (EFI): Measured electric fields and plasma waves, giving information on the behavior of charged particles in the magnetosphere.
- Fluxgate Magnetometer (FGM): Measured magnetic fields with great accuracy, allowing for the investigation of magnetospheric substorms.
- Electrostatic Analyzer (ESA): Analyzes ion and electron distributions to provide a thorough image of particle populations in space.

- The Solid State Telescope (SST) detected high-energy particles, which helped researchers comprehend particle acceleration mechanisms.
- Search Coil Magnetometer (SCM): Measures magnetic fluctuations, which are necessary for researching wave-particle interactions.

2. Design & Propulsion:

- Weight: Around 130 kilos apiece.
- Power Source: Solar panels providing around 50 watts.
- Propulsion System: Hydrazine thrusters for orbital adjustment and station-keeping.

3. **Communication and data management:**

- Data Transmission: Telemetry and command communications are handled using an S-band radio.
- Onboard Storage: Solid-state recorders with a 1 GB capacity to ensure data integrity between transmission periods.

ARTEMIS satellites

The ARTEMIS mission repurposed two of the THEMIS probes, P1 and P2, for a new mission to investigate the lunar space environment. This repurposing necessitated considerable changes to the satellites to meet their expanded mission goals.

1) Instruments:

- Combined Instrumentation: The whole set of THEMIS instruments was retained, ensuring continuous data gathering and analysis.
- Lunar Dust Detector: Added to determine the effect of lunar dust on space weather occurrences.

2. Lunar-Orientated Design:

- Orbital Changes: Moved from Earth-centric orbits to lunar Lagrange points and low lunar orbits.
- Thermal Control: Improved thermal control technologies to deal with high temperature fluctuations near the moon.

3. Communication & Navigation:

- Extended Range: Improved communication technologies to accommodate longer distances from Earth.
- Autonomous Operations: Created onboard algorithms to boost autonomy, which is critical for operations away from Earth-based control.

The move from THEMIS to ARTEMIS demonstrated NASA engineers and scientists' creativity in transforming existing assets into cutting-edge lunar exploration instruments. The effective adaption of these satellites demonstrates the versatility and inventiveness required for current space research.

Appendix B: Key Data and Findings from the THEMIS and Artemis Missions

The THEMIS and ARTEMIS missions provided a plethora of scientific data that greatly aided our knowledge of space weather occurrences. This appendix highlights the important discoveries from these missions, highlighting their significance to our understanding of the Earth's magnetosphere and the lunar space environment.

Key Findings from THEMIS

1. Magnetospheric Substorm:

THEMIS made the first extensive observations of substorms, showing the time and geographical dynamics of these occurrences. The data showed that substorms are caused by the abrupt release of energy held in the Earth's magnetotail, which results in auroral displays and increased particle acceleration.

2. Plasma Sheet Dynamics:

The expedition revealed the intricate dynamics inside the plasma sheet, demonstrating how magnetic reconnection events cause particle energization and transport. These ideas have been essential in understanding how energy moves through the magnetosphere.

3. Solar Wind Interaction:

THEMIS demonstrated the complicated link between the solar wind and the Earth's magnetosphere. The data showed how fluctuations in solar wind pressure and magnetic field direction affect the magnetospheric boundary, resulting in phenomena like magnetic storms and geomagnetic pulsations.

Key Findings of ARTEMIS

1. Lunar Wake Dynamics:

ARTEMIS witnessed the production of a "lunar wake" when the moon interacts with the solar wind, shedding light on the behavior of plasma around airless bodies.

The expedition revealed how the wake alters charged particle distributions and magnetic field structures.

2. Magnetic Tail and Lunar Environment:

The mission discovered that the moon serves as a natural laboratory for investigating the Earth's magnetotail. The ARTEMIS data revealed how the lunar environment reacts to fluctuations in the magnetotail, providing a unique viewpoint on magnetospheric processes.

3. Electrical Fields and Plasma Waves:

ARTEMIS detected electric fields and plasma waves in the lunar environment, emphasizing their effects on particle mobility and energy transmission. These discoveries are important for understanding how space weather impacts airless entities and their surfaces.

4. Water, Volatiles:

ARTEMIS gave indirect evidence of water and volatile transport throughout the lunar surface. The mission's discoveries revealed that solar wind interactions might play a role in the transport and sequestration of these critical materials, opening up new paths for future research and use.

The THEMIS and ARTEMIS missions have transformed our knowledge of space weather, revealing new details about the dynamic interactions between celestial bodies and their surroundings. The data obtained by these missions continues to educate and inspire future study, ensuring that their effect is seen for years to come.

Appendix C: Glossary of Terms and Acronyms.

This glossary defines important words and acronyms used throughout the book, giving readers a fast reference to help them comprehend the ideas and terminology associated with the THEMIS and ARTEMIS missions.

- Aurora: A natural light show in the sky created by the collision of energetic particles with the atmosphere, which is most often observed in high-latitude locations.
- Electrostatic Analyzer (ESA): A device that monitors the dispersion of ions and electrons in a plasma.
- Fluxgate Magnetometer (FGM): A device for measuring the intensity and direction of magnetic fields.
- A geomagnetic storm is a transient disruption of the Earth's magnetosphere generated by solar wind shock waves and magnetic field disturbances.
- Hydrazine Thrusters: A chemical propulsion device used to move spacecraft in orbit.
- Lagrange Point: A location in space where the gravitational forces of two big things,

such as the Earth and the Moon, balance the centripetal force experienced by a smaller object.

- Magnetospheric Substorm: A transient disruption in the Earth's magnetosphere that releases energy and increases auroral activity.

- Magnetotail: An extended extension of a planet's magnetosphere on the opposite side from the sun, formed by the solar wind.

- Plasma Sheet: A area of the magnetosphere that contains high-energy charged particles and is positioned between the magnetotail lobes.

- Solar Wind: A stream of charged particles (mostly electrons and protons) emitted from the sun's upper atmosphere.

- Solid State Telescope (SST): A device used to detect and monitor high-energy particles in space.

- Time History of Events and Macroscale Interactions During Substorms (THEMIS): A NASA project that investigates the Earth's magnetosphere and its interactions with the solar wind.

Appendix D: A list of contributors and collaborators

The THEMIS and ARTEMIS missions were successful because to the devotion and competence of a varied team of scientists, engineers, and researchers. This appendix recognizes the major contributors and partners who played critical roles in the missions, emphasizing their efforts and accomplishments.

NASA Leadership and Management:

- Dr. Vassilis Angelopoulos is the Principal Investigator for THEMIS and ARTEMIS, responsible for scientific goals and mission strategy.
- Dr. David Sibeck: Project Scientist, in charge of coordinating scientific research and data analysis initiatives.
- Dr. Jim Burch is a co-investigator who focuses on instrumentation development and mission operations.

Engineering and Technical Teams:

- Goddard Space Flight Center: Contributed engineering knowledge to spacecraft design, integration, and testing.
- Jet Propulsion Laboratory (JPL): Supported mission planning, navigation, and communication systems.

International collaborators:

- The Canadian Space Agency (CSA) provided equipment and scientific expertise for magnetospheric research.
- European Space Agency (ESA): Worked together on data processing and scientific research linked to lunar exploration.

Scientific Community:

- Dr. Margaret Kivelson directed research on magnetospheric dynamics and wave-particle interactions.
- Dr. Christopher Russell made contributions to the study of magnetic field phenomena and solar wind interactions.

These people and organizations worked together to ensure the success of the THEMIS and ARTEMIS missions. Their commitment to scientific discovery and technical innovation has aided our knowledge of space weather and its consequences for future missions.

Conclusion

The appendices of "ARTEMIS Unveiled: The Moon's Electric Shield" provide a complete description of the technical, scientific, and collaborative elements of the THEMIS.

www.ingramcontent.com/pod-product-compliance
Lightning Source LLC
Chambersburg PA
CBHW071924210526
45479CB00002B/544